NETWORKS

of

NEW YORK

INGRID BURRINGTON

NETWORKS

of

NEW YORK

AN ILLUSTRATED FIELD
GUIDE TO URBAN
INTERNET INFRASTRUCTURE

MELVILLE HOUSE
BROOKLYN • LONDON

NETWORKS OF NEW YORK

First Melville House Printing: August 2016

Melville House Publishing

Melville House Publishing		8 Blackstock Mews
46 John Street	and	Islington
Brooklyn, NY 11201		London N4 2BT

mhpbooks.com   facebook.com/mhpbooks   @melvillehouse

Library of Congress Cataloging-in-Publication Data
Names: Burrington, Ingrid, author.
Title: Networks of New York : an illustrated field guide to urban
    Internet infrastructure / Ingrid Burrington.
Description: Brooklyn : Melville House, [2016]
Identifiers: LCCN 2016022375 (print) | LCCN 2016022738
    (ebook) | ISBN 9781612195421 (hardcover) | ISBN
    9781612195438 (ebook)
Subjects: LCSH: Computer networks—New York (State)—
    New York—Equipment and supplies. | Internet—New York
    (State)—New York—Equipment and supplies. | Information
    superhighway—New York (State)—New York. | New York
    (N.Y.)—Buildings, structures, etc.
Classification: LCC TK5103 .B87 2016 (print) | LCC TK5103
    (ebook) | DDC 004.67/8097471—dc23
LC record available at https://lccn.loc.gov/2016022375

Design by Marina Drukman

Printed in China

1   3   5   7   9   10   8   6   4   2

CONTENTS

INTRODUCTION

"How do you see the Internet?"

Over the past two years, I've asked a lot of people this question. It's a question often met with confusion or requests for clarification. Do I mean "What do you *think* about the Internet, like in the grand scheme of things?" or "How do you think the Internet *works?*" or "How do you *access* the Internet?" Really, I'm asking all three.

Sometimes it helps to start with that last question: how people access or use the Internet. For most people, the answer is that they see the Internet through screens—browsers and apps on laptops and phones. Sometimes people will point at a router, vaguely understanding that's the device their Wi-Fi connection comes from.

Once I understand the specifics of how someone interfaces with the Internet, I'll ask the second question: "How do you think the Internet works, and how do you visualize that process?" At this, answers vary, though they tend to follow three common trajectories, each one aligning well to certain tropes seen in stock photos and illustrations sometimes used to describe "the Internet."

"I have no idea. Maybe black magic."

This answer sometimes involves hand-waving and anxious, slightly apologetic faces. To be fair, there are a lot of bad stock images out there that support that answer—which is to say, they make the mechanisms of the Internet appear impossibly complicated and opaque. You may have seen these sorts of images. Sometimes it's a man at the peak of an over-Photoshopped mountain, his arms reaching for a giant laptop in the sky from which fluffy white clouds emerge. Other times, it's a different man (always men in these weird dreamscapes, usually wearing ties), hands cradling poorly rendered collages of computers and a globe floating in some ethereal mist that could be data traveling across the network—or could be fairies, no one really knows.

Ironically, some of these baffling images emerge from attempts to make the Internet seem *less* complicated, through metaphors like "the cloud." Metaphors can be useful teaching tools, but when *all* that people know about the Internet are metaphors, it tends to make their understanding of it more clouded, not less.

This is the most pragmatic answer, and it's the one accompanied by stock images of network diagrams—icons of computers in dots, connected to one another with lines in an ever-expanding web. And it's slightly more accurate, but it's still an abstraction. It doesn't really convey the scale of the network, the volume of objects that are part of it, or the places where nodes in that network end up being concentrated. In reality, the Internet isn't an evenly distributed utility—nodes aren't all perfectly connected, and the quality of a connection can vary widely. While abstract network diagrams are good for getting a general idea of how it's supposed to work, they sweep up the particularities of geography and history that play a large role in shaping the Internet.

"There's a whole bunch of . . . computers somewhere doing . . . stuff. Cyber stuff."

We're getting warmer with this answer. There's actual reference to *objects*. We don't necessarily understand what they're *doing*, and maybe don't understand how they work

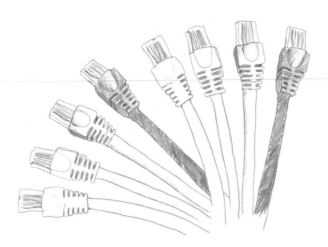

together, but they *exist*. The visuals associated with this answer tend to be extreme close-ups of cables and Ethernet ports or wide shots of dark hallways of server racks, illuminated only by the softly pulsing blue LED lights of machines. Rooms of wall-to-wall screens with incomprehensible diagrams or CCTV footage streaming across them.

There's an implicit assumption in these images that the Internet's physical spaces are privileged spaces, behind locked doors with retina scanners and high gates. These security procedures make sense, but the theater and optics that surround them encourage an assumption that only an elite few are clever and privileged enough to understand how the Internet actually works—and if you aren't one of those people, being curious about how the Internet works is circumspect.

. . . .

The thing is, all of these answers—that the Internet is made up of objects, that it's a network, that it's weird black magic—are kind of correct. The magic part isn't necessarily how the machines work but rather the weird, amazing, terrifying, and wonderful things that people figure out how to do with the network.

These notions of how the Internet works, and the images that illustrate them, are how I saw the Internet two years ago when I started asking people that annoying question. I didn't really know what the Internet looked like, but I was pretty sure it didn't look like any of the images I saw paired with headlines about Edward Snowden or Facebook or Google. I wanted to know about the stuff that made the Internet work, where it lived, and how I could find it. There are some excellent public resources on this topic, but for some questions I really wanted to talk to someone on the inside. However, with a few exceptions, almost everyone "official" I tried to talk to brushed me off. The rejections didn't read as a matter of security or of protecting the networks but more like a failed transaction. I wasn't specialized, I wasn't already an expert, I wasn't well connected, and I couldn't get anyone good press with my work. I was just some curious artist, so nobody really had any interest in answering my questions. I had to find the Internet on my own, starting where I already was: New York City, a weird set of islands off the coast of America.

This is how I find the Internet, now that it's been two years since I started looking: I go for a walk literally anywhere in the city. In the span of about three city blocks, I can usually figure out where there's buried fiber optic cable (and, sometimes, who owns that cable), which devices hanging above traffic intersections are talking to one another, how many cell towers are in the area, and whether I'm currently under surveillance (by the NYPD or a private observer).

When I do this, I don't have a cool gadget that shows me radio frequencies moving through the air or some kind of Internet divining rod. I have no secret expertise or ninja hacker skills. I'm just looking down at the ground and looking up at buildings. That's it. To me, this is another kind of Internet magic—maybe a more practical magic, the kind that recognizes the existence of things larger than oneself and one's small worldview. It's weirdly comforting, inspiring even, to walk down Sixth Avenue with the knowledge that buried underneath my feet is a fiber optic cable that is carrying conversations, photographs, stories, secrets, and *lives* as beams of light through hair-thin strands of glass. *Networks of New York* is a guide for practicing the everyday magic of seeing the Internet as part of a city's landscape and everyday life.

It's also a guide to some of the weird history and politics that shape how the Internet weaves its way through the city. In the United States, communications networks tend to be built on top of existing networks—fiber optic cable routes follow telephone routes, which in turn follow telegraph routes. This makes a lot of sense—it's easier to build on top of systems that already work than to start entirely from scratch. But it means that in especially old American cities (like New York, Chicago, San Francisco), the city's networks exist atop the sediment of past technologies. It's why buildings once crucial to the telegraph system are now home to hundreds of thousands of miles of fiber optic cable, and why those buildings' rooftops are now adorned with satellite receivers and microwave antennae.

Networks also tend to inherit the legacies of past networks in the same way that they inherit landscape. With each further compression of time and space through communications technologies, there's a lot of hopeful zeal that this will be society's turning point—that as radio waves and cables connect us across greater and greater distances, people will be closer, with greater freedom and greater choice than ever before. But that hopeful promise associated with an expanding Internet depends on equal access to it, and as examples in this field guide demonstrate, network access isn't necessarily evenly distributed to everyone. When the Internet is mostly owned by private companies, there's little financial incentive to treat it as a public right. The increasing presence and complexity of surveillance systems throughout cities also mean that networked technologies limit freedom as much as they enable it, and in places like New York, those systems tend to specifically observe and target minority populations.

New York's networks and the politics of those networks in some ways reflect very local concerns. Banks fighting for server space in data centers to shave microseconds off a financial transaction isn't necessarily every American city's Internet story. But some of New York's Internet stories—private companies monopolizing consumer Internet access, post-9/11 surveillance boondoggles, and a growing enthusiasm for data-driven cities leading to more networked sensors and objects in public space—do translate to the concerns and infrastructures of other cities. And while some of the specific companies and types of sensors and cameras mentioned in this field guide are distinct to New York, the *types* of objects documented (manhole covers, excavation markings, antennae, cameras) can be found in pretty much any American city. It's actually one of my favorite parts of visiting different U.S. cities. While some people seek out local cuisine or important landmarks, I get excited over spotting interesting telecom manhole covers.

One of the hardest parts of trying to see the Internet, of trying to even answer the question of how you see the Internet, is scale. The writer Quinn Norton has written about the difficulty of telling stories today in "a world where falling in love, going to war and filling out tax forms looks the same; it looks like typing." There is an unexpected intimacy to living with screens, but that intimacy does not typically extend to the cables and conduits the screens rely upon. As the division between "real life" and "online life" is increasingly understood to be a fiction (i.e., what people say and do online has real-world consequences, retweets are not endorsements, your boss can find your Tinder profile), the Internet's landscapes continue to appear at a remove from those physical landscapes in which we fall in love, go to war, and fill out tax forms. Ironically, the reason we can even have those weirdly personal moments with machines is because the landscapes of the Internet are folded into the landscapes of everyday life. We basically live inside a really big computer.

This is not an exaggeration. To understand this point, it's helpful to look at how people used to live with computers. When discussing this topic, the artist James Bridle often cites a quote from mathematician Harry Reed about the Electronic Numerical Integrator And Computer (ENIAC), one of the first modern computers, created during World War II: "The ENIAC itself, strangely, was a very personal computer. Now we think of a personal computer as one which you carry around with you. The ENIAC was actually one that you kind of lived inside."

While the hardware that we use to connect with one another has gotten a lot smaller, I don't think that computers have really become smaller since the days of the ENIAC. It's more that the room has just gotten bigger. It's not really a room so much as it's a planet. What we think of as personal computers today are just bits of aggregate hardware in a much larger, more complicated computer that is the Internet. But living inside a computer doesn't look like a cool science fiction movie or any of the stock images used to describe the Internet. It looks like cities, highways, buildings, and the infrastructure that supports them.

Infrastructure is the stuff of cities that only really gets noticed when it stops working. It's not so much invisible as it is hard to see; it hides in plain sight. The trick of how to see the Internet isn't tech know-how or gaining access to secret rooms. It's learning what to look for and how to look for it. Learning how to see and pay attention to the fragmented indicators and nodes of networks on any city street is also a process of learning how to see and live within a world full of large, complicated systems. These systems often feel at a far remove from everyday life but frequently (and, in the case of the Internet, increasingly) shape and transform it. Hopefully this field guide will help you navigate those systems as much as it will help you navigate New York City.

BELOW THE GROUND

When trying to find Internet infrastructure in a city, it's helpful to start from the ground up—or, more precisely, somewhere just below the ground up, under manhole covers, inside underground ducts, and even farther below on subway platforms. Because getting online today usually involves connecting to a wireless network, it can be easy to forget that most of the Internet is, in fact, a series of tubes; wireless connections still have to go through wires. For example, when you use a Wi-Fi network in a public library, your Internet activity travels wirelessly to a router. That router is wired into a cable network, usually with an Ethernet cable. And that cable sends your Internet activity back into a global network made mostly of fiber optic cable. At the end of the day, most of the things we read and do online are reduced to pulses of light in glass tubes. And in New York, many of those glass tubes are *just underneath your feet*!

Despite being so close beneath our feet, it's not exactly easy to find those glass tubes. They're mostly buried. Internet cables in New York live in utility ducts, which are owned and maintained by three companies, Empire City Subway (ECS; no relation to the transportation subway system), Verizon New York, and Consolidated Edison Company of New York (also known as Con Ed, the local power company). Different companies own cables (copper, coaxial, or fiber) that run through these ducts. To run cables in the ducts, companies need to acquire a franchise from the city's Department of Information Technology and Telecommunications (DoITT). The agreements stipulate that the companies pay the city a percentage of their revenues in New York City, and that the company has to provide an amount of dedicated fiber to the city.

It's pretty hard to find a telecommunications company or city utility company that will just give you a map of where all the fiber optic cables are located. Usually the rationale for this is security, but it's also because private companies view their networks as essentially proprietary business information, i.e., trade secrets. Still, it's possible to find the Internet under your feet without the aid of corporate maps— it's mostly a matter of knowing what to look for.

STREET MARKINGS

Sometimes while walking in a city, you might notice colorful markings on the street—neon arrows and labels, usually barely legible. The markings extend throughout the street, sometimes in zigzagging directions. This urban markdown language isn't really intended for people on the street. Whenever a contractor or construction company plans to do street excavation, they have to call 811, the nationwide number for utility-locating services. The contractor gives 811 information on where they'll be excavating, and 811 in turn contacts local utility companies— gas, electric, water, and telecom—to alert them that they need to mark out the location of their underground cables so that the contractor knows to watch out for them. In New York, people who locate utilities tend to use spray paint to mark out buried ducts; in more rural areas they'll sometimes use tiny flags or install signposts.

There's a standardized color code for street markings recommended by the American Public Works Association (APWA) and used by many cities, including New York (see inside back cover). Orange refers to the broad catch-all of "communications, alarm, signal lines, cables, and conduit." This means that orange lines represent Internet cables, television cables, or telephone lines—literal signs of the circulatory systems of the Internet and all the other networks around us.

These spray-painted markings aren't placed on the ground for the sake of a curious public but for construction crews, and their aesthetics reflect this target audience. They're often sloppily written with letters like "ECS" reduced to a single looping scribble. On rare occasions I've seen markings that include question marks, as if someone didn't entirely trust the map they used to locate buried utilities. The markings, like the construction sites they're created for, are also temporary. Depending on the weather, how much traffic crosses an intersection, and further excavation work, they fade away over the course of either a few months or a few weeks. Sometimes several different markings will cut across the same intersection, turning city streets into haphazard tapestries of excavation history.

While these markings aren't meant to be legible to the general public, they do offer a useful way of reverse engineering the locations of buried communications networks in a city. The level of detail about these networks that can be gleaned from the markings varies, as we'll see with the following examples—sometimes they indicate only that *some* kind of communications duct is underground, sometimes they indicate the type of cable underground, and sometimes they indicate what company owns that cable. The company names, however, can be a bit misleading—as we'll see, most telecommunications companies are not so much companies as they are chimeric creatures resulting from decades of companies forming, merging, and spinning off from or acquiring other companies. For the reader who might find the following histories of company mergers and acquisitions tedious, it is helpful to replace "company" with "dragon" and words like "acquired" with "devoured." It is not an inaccurate image for what mergers and acquisitions are really like.

1. Types of Cable

FIBER OPTIC CABLE

"Fiber optic" doesn't necessarily mean what's commonly thought of as the Internet—a variety of data types (Internet, video, voice communications) can run over a fiber optic cable. It also might not be part of what's considered the *public* Internet. Private companies sometimes maintain their own networks for internal communications, as do city agencies and universities.

CABLE TELEVISION

This means pretty much what you think it means: a cable for television, usually coaxial (for more, see page 98).

TELEPHONE LINES

Traditional land-line telephone connections are composed of two wires, historically made of copper, which run to the nearest telephone exchange building. While many have abandoned land lines for mobile phones and other online telephony services (known as Voice over Internet Protocol or VoIP), many land lines remain buried underground and in use.

TELEVISION MARKING

This cable is more commonly seen in boroughs other than Manhattan, for reasons that remain unclear from field observation. Although the labeling indicates a television cable, most consumer Internet Service Providers (ISP) in New York operate as "bundlers" selling television, phone, and Internet connections in one package, and all three services are often transmitted along the same cables.

2. Characteristics of a Duct

DUCT WIDTH MARKINGS

These markings, which may look like cave painting in-terpretations of TIE fighters in *Star Wars*, are standard markings approved by the APWA for marking the width of belowground ducts.

SHALLOW CABLE

This means basically what it sounds like: the under-ground cable isn't very deep under the pavement. This marking is a warning to anyone doing street excavation work that they might hit the cable. Another easy way to identify these shallowly buried ducts is to look at the road itself. Is the road a uniform color or does it have splotches of darker gray or black? Do the splotches run in uniform, parallel lines, about eighteen inches apart? These are most likely patches placed on the road after paving over a new, shallowly buried duct.

This refers to the point at which a cable duct enters a building, typically entering at basement level. Usually these are seen at the edge of a building and not on the road itself.

3. Companies

●

EMPIRE CITY SUBWAY

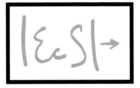

Empire City Subway was formed in 1891 to construct and maintain tubes for telegraph and telephone cables ("subway" referred to anything underground, not just transportation networks) in Manhattan and the Bronx. Prior to that, all cables—communications and electric—were strung across the skyline on poles because it was cheaper for companies to keep adding cables above the ground than deal with the construction costs of putting them underground. Illustrations of Lower Manhattan in the 1880s depict the area under the shadow of densely layered cables, which sometimes snapped in wind- or snowstorms, lashing electric current through narrow city streets. It took the Great Blizzard of 1888 and the resulting massive property damage to finally push utility companies to bury cables underground.

ECS operates only in Manhattan and the Bronx because in 1891 New York City was basically just those two boroughs. ECS now leases space in its conduits to telecommunications companies. They own approximately 11,000 manholes and 58 million feet of conduit. A marking that indicates "ECS" doesn't provide a ton of insight into the types of cables in a duct or who might own those cables; mostly it's just an indicator that *some* type of telecommunications network is in the area.

AT&T

AT&T has been a presence in New York City almost as long as telephones have have existed in New York City. The New York–based American Telephone and Telegraph Company was initially formed in 1885 by executives from Alexander Graham Bell's American Bell Telephone Company to create a nationwide long-distance network. AT&T first connected Chicago and New York in 1892 and then acquired American Bell, the company that technically created AT&T, in 1899. Corporate offspring devouring its parent company isn't that unusual an origin story, but it is unusual that in AT&T's case, it has happened more than once. After decades of holding a monopoly on the U.S. telecommunications industry,

AT&T spun off its regional companies into seven separate entities in 1982 as part of an antitrust settlement with the U.S. Department of Justice. One of those spun-off companies, Southwest Bell, ended up growing to be one of the largest telecommunications companies in the United States, in part by taking advantage of changes to Federal Communications Commission rules in the mid-1990s. In 2005, what had been Southwest Bell (rebranded as SBC Communications in 1995) acquired AT&T and took on its parent company's branding.

While this field guide might not be the place for the entirety of AT&T and the Bell System's voluminous history, traces of that history can be found all over New York City. The company initially formed to create long-distance networks that would become known as AT&T's Long Lines division, and there are Long Lines switching stations and offices throughout Manhattan—a particularly Brutalist monstrosity at 33 Thomas Street, another switching center at 811 Tenth Avenue, and the former central Long Lines office at 32 Avenue of the Americas, which is now a major convergence and colocation center for networks.

LEVEL 3 COMMUNICATIONS

Level 3 Communications began trading on NASDAQ in 1998 and received its franchise to build a fiber optic network in New York City in 1999. They are a major Tier 1 network, which means that their network has a direct

connection to every other network online without paying fees to do so—a process known in the telecom world as peering (for more on Level 3, see its Manhole Covers entry; for more on peering, see Glossary of Terms).

LIGHTOWER

Emerging in 2006 from the ashes of what once was National Grid Wireless US, Lightower established itself as a purveyor of dark fiber (see Glossary of Terms) in New York City through a series of company acquisitions. Two of the crucial network expansions were in 2013, through a merger with Sidera Networks (formerly a part of the cable company RCN; for more, see RCN later in this section) and the acquisition of Lexent Metro Connect (formerly owned by the owners of Hugh O'Kane Electric, one of the most important cable splicing and fiber pulling companies in New York; for more, see Ground Level).

Lightower is majority owned by Berkshire Partners, a Boston-based private equity firm that has invested heavily in other telecommunications infrastructure companies, such as colocation services company Telx (for more on colocation, see Carrier Hotels in Ground Level), mobile infrastructure company Crown Castle International (for more, see Distributed Antenna Systems in Above Ground), and a variety of wireless antenna tower companies operating throughout the Americas and southeast Asia.

MCI

When the company began in 1963 with the intention of building out a microwave transmission network between Chicago and St. Louis, MCI stood for Microwave Communications, Inc. And like many entries in this field guide, it was devoured and transmogrified many times over by the dark alchemy of corporate mergers and acquisitions. Known as MCI World-Com and later WorldCom between 1998 and 2003, the corporation reclaimed its original name following WorldCom's spectacular bankruptcy and corresponding financial scandals in 2002. While MCI as a company no longer exists (it was acquired by Verizon in 2006), its name has apparently not been completely phased out of use in network maps in New York.

NEXTG NETWORKS/CROWN CASTLE

In 2008, NextG Networks received a city franchise agreement to install wireless infrastructure on city street poles for its distributed antenna system (DAS—for more, see Distributed Antenna System in Ground Level). The agreement also permits the company to install fiber belowground "for purposes of connecting Base Stations installed on Street Poles to one another or to a supporting telecommunications system." These fiber markings are often found near street poles or at the bases of street poles. In 2011, NextG was acquired by mobile infrastructure company Crown Castle International. Both names tend to appear on their street markings, usually seen at the base of a street post with a DAS attached.

RCN

RCN is a major Internet service and television cable provider, but it's unclear whether a cable marked "RCN" is actually a cable that belongs to RCN or a cable that once belonged to a division of RCN. As many other examples in this field guide demonstrate, while cable networks change corporate parents and change names pretty frequently (particularly in the early 2000s), the maps used by utility locators aren't updated accordingly.

In 2001, Con Edison announced that they were creating their own fiber optic network, managed by the holding company Con Edison Communications. In 2006, RCN acquired Con Edison Communications for $32 million. Private equity firm ABRY Partners acquired RCN in 2010. The firm subsequently repackaged and rebranded its metro area operations (which managed the former Con Ed fiber network) as Sidera Networks. Sidera merged with Lightower Networks in 2013 (after both companies were acquired by Berkshire Partners in 2012). Both Lightower and RCN operate in New York, but given how many dated company names appear in sidewalk markings (see MCI), it's conceivable that an RCN marking is, in fact, now a Lightower cable.

SPRINT

Sprint, a company that emerged from the U.S. railroad network (Sprint is actually an acronym for Southern Pacific Railroad Internal Networking and Telephony), was one of the more forward-thinking telecommunications companies. It built the first nationwide fiber optic cable network in 1986. Today, the cables it runs throughout the city are primarily for network traffic on mobile phones (for more, see Cell Towers in Above Ground).

TRANSIT WIRELESS

Transit Wireless provides wireless connectivity to the Metropolitan Transportation Authority's subway platforms (for more, see Subway Wireless Networks).

WINDSTREAM COMMUNICATIONS

While at the time of this writing, Windstream markings have only been sighted at 60 Hudson Street, a major

New York carrier hotel (for more, see Carrier Hotels in Ground Level), the Little Rock–based company is worth noting here due to a clever trick it pulled in the summer of 2014 that could reflect future patterns in telecommunications businesses. In 2014, the IRS and the U.S. Treasury expanded "eligible assets" for corporate entities known as real estate investment trusts (REIT) to include cables, transmission devices, and pipelines. Traditionally, REITs are companies that own or finance actual properties—houses, office buildings, land. The idea that infrastructure could constitute "property" in the same sense as a building is in and of itself philosophically interesting, and it also offers a useful financial loophole for companies that own a lot of infrastructure, since REITs are essentially tax-exempt entities. After the rule change went into effect, Windstream spun off its copper and fiber network into a REIT. The new REIT "leases" its property (i.e., the network) to Windstream for around $650 million, which Windstream can write off as a business expense. Basically, it is a way of funneling a lot of money away from taxes and into a black hole. Presumably Windstream markings in Lower Manhattan technically belong to this REIT.

XO COMMUNICATIONS

In 2002, like many network providers that built fiber optic networks in the mid-1990s, XO filed for bankruptcy. It emerged from bankruptcy in 2003 with well-known activist shareholder Carl Icahn as its majority shareholder and board chairman. In 2008, Icahn orchestrated a massive refinancing of XO's debt (90 percent of which Icahn owns) and took the company private in 2011. A lawsuit brought in 2009 by XO's minority shareholders, which accused Icahn of using XO's financial losses as tax benefits for other companies he owns, remains ongoing as of 2016.

MICROTRENCHING

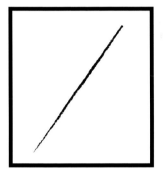

Microtrenching is a technique for installing conduits into narrow "slot-cut" trenches in the ground. Currently the only fiber lines installed with microtrenching in New York City are owned by Verizon, who received a franchise for a microtrenching pilot program in 2013 as part of its franchise agreement to provide citywide fiber optic connectivity (for more, see Verizon sidebar).

Microtrenching involves less intensive excavation work to install than going deep into underground ducts, but only a limited number of fiber strands can be added and they are less protected from the elements. Signs of microtrenching resemble scars: places where it looks like narrow cuts have been made, and then repaired, on the street.

Verizon: The Infrastructural Elephant in the Room (or Really, in the Ducts)

While spray-paint markings or manhole covers for Verizon have yet to be sighted in the wild in New York, the telecommunications company has in many ways left its mark on the city's Internet infrastructure. Technically, it owns the majority of it. Verizon is the result of a 2000 merger between telecommunications giants GTE Corporation and Bell Atlantic, which included inheriting both Empire City Subway and the New York Telephone Company (now Verizon New York), the companies that own and maintain telecommunications ducts beneath New York City. In the past decade, however, Verizon has received more attention from the city for its often stalled efforts to implement new infrastructure, not for the infrastructure its subsidiaries maintain.

In 2008, Verizon entered into a franchise agreement with DoITT to build out an all-fiber network (branded Fios by Verizon) to serve all New York City residents with access to fiber optic cable connections. At the time, this was seen as a major opportunity to address the city's digital divides by bringing high-speed broadband to underserved areas still running connections across old coaxial or copper lines. According to the terms of the agreement, once a part of the Fios network "passed" (a distinction that will become relevant later) a building, that building's owner could request Fios service be installed, at which point Verizon would have a six-month window to do so. If Verizon's installation was obstructed, it would be required to explain the reason for the delay in writing and resolve the matter within another six months (a scenario called a "non-standard installation" or NSI). If Verizon was unable to bring fiber into a multiple-dwelling building (i.e., an apartment building), the company could petition the state Public Service Commission for something called an Order of Entry Petition. Procedures for NSIs and multiple-dwelling units were written into the franchise agreement.

The contract required the company to "pass" all homes in every borough by the end of June 2014. While Verizon claimed in November 2014 that it had successfully "passed" all New York City households, as of 2016 many New Yorkers remain disconnected from Fios and unable to get it. An audit published by DoITT in June 2015 (based on limited access to Verizon's apparently very sloppy internal records) indicated that almost 75 percent of the NSIs outstanding at the time of the audit had been unresolved for longer than the twelve-month limit. While mostly anecdotal, reporting on the Fios rollout suggests that Verizon concentrated primarily on new developments and already wealthy or gentrifying neighborhoods, bypassing, and at times outright refusing, to provide service to other neighborhoods or buildings with no reason given.

Verizon and its lawyers argue that "connecting" was never the actual objective, leaning heavily on the fact that the original franchise agreement stated "passing" without having clarified if a cable passing a building has to be *accessible* to that building. They also argue that gaining access to multiple-dwelling units, dealing with landlords, and the costs and logistics of pulling new fiber into buildings is really, really hard. (Their response to DoITT's audit is a magnificent demonstration of the hypersensitivity of corporations to the conditions of reality.)

While this aside on Verizon's poorly implemented fiber network will admittedly quickly become dated, it's included as a reminder of the inherent obstacles to attempting city-wide connectivity improvements. But difficulties aside, Verizon has little incentive to negotiate these tricky, potentially low-return-on-investment installations. The franchise agreement doesn't significantly penalize Verizon for failing to meet its terms.

This isn't to say citywide fiber connectivity is impossible; other cities have done it. In Chattanooga, Tennessee, for example, the city's electric company EPB added a citywide gigabit fiber network parallel to the power grid. Since all Chattanooga households were already connected to the electricity network, the process of pulling in new fiber was far simpler than Verizon's debacle. But smaller cities can more easily get away with building out municipal Internet services. In larger cities like New York, where telecoms have very established footprints, private-sector solutions are typically favored over public ones.

MANHOLE COVERS AND HANDHOLES

Once you've spent enough time looking for and at cable markings, you'll notice that the markings tend to go into or emerge from manholes on the street. Manhole covers are a literal point of entry into the city's underground world and their designs indicate what part of that world they connect to (e.g., the power grid, the gas system, or the telecommunications network). While the duct networks still belong to the aforementioned ECS, Verizon New York, and Con Edison, and most telecom duct manhole covers feature the same design, one aspect of having a franchise agreement in New York is that it entitles the franchise holder to install their own manholes and manhole covers in areas where they own a significant amount of cable.

The New York Telephone Company (also known as Bell Atlantic New York) was ECS's original primary shareholder, and the hexagonal pattern on ECS manhole covers can be seen in Brooklyn, Queens, and Staten Island with the Bell logo in the center rather than "ECS." Through the dissolution and reconstruction of what was the New York Telephone Company (for more, see Verizon sidebar), Verizon inherited all of its holdings, which means ECS is now a wholly owned subsidiary of Verizon. Verizon also maintains the ducts underneath the boroughs that ECS doesn't cover, under the subsidiary company Verizon New York. Manholes outside of ECS territory have the same hexagonal pattern but feature the old Bell logo instead of the ECS logo.

Level 3's global fiber network is massive, but the company is not a household name. This is partly because it is kind of too big to bother with residential Internet services, so it instead focuses on providing services to businesses, hospitals, and, in some cases, governments. In 2012, Level 3 received a $411 million contract from the Department of Defense's Defense Information Systems Agency (DISA) to provide fiber cable and maintenance support to DoD networks. This is not a particularly remarkable thing for a network operator to do (lots of companies provide similar services to the DoD). However, it might explain why Level 3's Chelsea colocation facility is in the same building as the New York offices of the FBI's Joint Terrorism Task Force (JTTF). Or those things could be completely unrelated; as we'll see

in later sections of this guide, New York City real estate is home to many weird overlaps of law enforcement and Internet infrastructure. This makes some sense, as both require a lot of space and security and are quite good at quietly insinuating themselves into the fabric of cities.

Level 3's franchise agreement with the city dates back to 1999. In 2010, an audit from the city's comptroller office determined that Level 3 had underreported its revenue to the city for a number of years. The report called on the company to pay back $543,000 in fees and interest. In 2012, Level 3 board member Rahul Merchant was named commissioner of DoITT and the city's first citywide chief information and innovation officer. At this time, it is unclear how or if the Level 3 fees issue was resolved, although Merchant is no longer on the Level 3 board and left DoITT in 2014.

Time Warner as the Internet/television/phone services company we know came into existence in 1992 following the merger of Time Inc. and Warner Communications. They launched Time Warner Cable, one of the first high-speed cable Internet services, in 1996 before it separated from its parent company in 2009. Time Warner Cable's franchise agreement with the city of New York dates back to 1997, so presumably these manhole covers appeared sometime between 1997 and the present. Apparently that logo is supposed to be an eye merging with an ear, which is way more frightening than the hypnotic eye of Sauron that the casual observer might initially assume it represents.

TRANSIT WIRELESS

Transit Wireless was able to start placing their own manhole covers in Brooklyn in 2016 (for more, see Subway Wireless Networks).

MYSTERY MANHOLES

There is a degree of hubris to the custom manhole, a hubris that assumes a business model (not to mention a corporate logo) is as enduring as a slab of steel. Though less common in New York, in many American cities telecom manhole covers are portals into their network history—the logos of now-defunct companies like Qwest and Global Crossing still live on Chicago and Minneapolis streets. These manholes don't really persist out of nostalgia but rather laziness. The cost of replacing manhole covers featuring nonexistent companies with the manhole covers of the companies that have replaced them is a lot higher than whatever gains in brand recognition the replacements might win a company.

While the inevitability of corporate mortality (or rebranding following a merger) doesn't seem to deter some companies from investing in their own manholes, for those who are less inclined to make the investment, there are generic manhole covers that merely identify the use for the cables below for some kind of "communication."

HANDHOLES

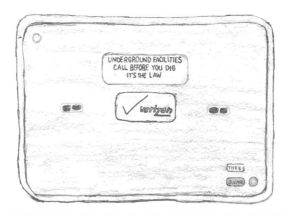

Handholes are structures set belowground that protect telecommunications cables and provide convenient access for splicing or pulling cable. The primary distinction between handholes and other manhole covers is that the former aren't as deep and don't tend to be as crowded as the ducts below manholes.

SUBWAY WIRELESS NETWORKS

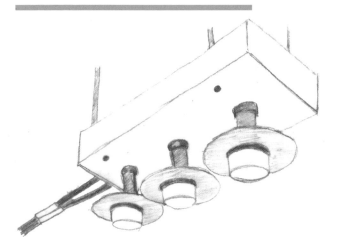

In 2005, the MTA issued a Request for Proposal (RFP) seeking a contractor to build out wireless services for Wi-Fi and mobile connectivity at subway stops. The company Transit Wireless formed in response to that RFP and, after receiving the contract, it formally began construction in 2007. As of 2015, the company is approximately halfway through its seven-phase building process and expected to finish connecting the entire subway system by 2017.

Transit Wireless's network is essentially a distributed antenna system (see Distributed Antenna Systems in Above Ground). The company has five "base station hotels" where mobile carriers and Wi-Fi providers connect their networking equipment to Transit Wireless's network. Each base station is located in an area strategically selected to be no more than twenty kilometers away from any of its connecting stations.

Signals from the base station travel via fiber optic cable to the connecting stations and are converted into radio signal via wireless antennae known as radio frequency (RF) nodes. At the same time, the nodes convert wireless signals from cell phones in the stations into optical signals, which get sent back to the base station via fiber. Each wireless node device has three nodes, each dedicated to a separate service: cellular service from mobile carriers, Wi-Fi services from private companies (e.g., Boingo Hotspot or Google), and public safety communications (e.g., 911 and MTA emergency services), a required feature of Transit Wireless's Metropolitan Transportation Authority contract.

....

Of Fiber and Finance: An Aside on High-Frequency Trading

The Financial District is a great place to look for spray-paint markings—there's often a lot of construction in the area and, as excavation markings reveal, a whole lot of cables. Finance has shaped New York in many ways, including its communication networks, and it shows on the street.

High-frequency trading (HFT) was a technological shift in the financial world that emerged around 2005. Broadly speaking, HFT uses complex algorithms to automate huge quantities of trades over very short periods of time, taking advantage of small differences in price across markets or between bidding and selling prices. These differences might be fractions of pennies, but by performing millions of trades at fractions of seconds, HFT firms are able to

make huge profits. In 2009, HFT accounted for 60 percent of all trades in U.S. stocks.

The pursuit of a microsecond advantage led to a lot of demand on Wall Street for low-latency networks, a term used to describe length of delay in data transmissions. Lower latency means less delay and faster trades. After apparently reaching the limits of mathematics for increasing speed, traders turned to physical proximity for lower latency. Data centers that housed stock exchanges offered expensive colocation services that placed a trading firm's servers closer to the exchange servers to improve latency (since the cable connecting the servers was shorter, data traveled a shorter distance and got to the server faster). New companies emerged, promoting ultra-low-latency networks by leasing private fiber lines. One company, Spread Networks, built an entirely new fiber optic network from Chicago to New York to be able to achieve—and charge hundreds of thousands of dollars for—a three-millisecond advantage.

Just like its trades, HFT has changed rapidly over a short period and probably will change even more by the time this book is published. There have been retreats into private "dark pools" run by banks, and the efficiency of algorithms has reduced the very volatility those algorithms gamed. At the same time, low-latency fiber is being rejected in favor of wireless microwave networks (as data traveling even slightly less than the speed of light in a fiber optic cable isn't quite fast enough).

As the black boxes of finance become increasingly opaque, it's weirdly reassuring to be able to identify their limited physical traces in downtown Manhattan—although, it turns out, most of the major exchanges and banks now choose to keep their servers in nearby New Jersey.

GROUND LEVEL

If staring at things on the ground isn't your thing (or you've realized the hazard of only looking down while crossing streets in New York), there are plenty of surface-level indicators of network infrastructure. They're usually not as colorful as street markings, but they're just as ubiquitous, if not more so, since they tend to be more permanent than spray paint. Many of them are wireless devices, sending signals through the air and relaying signals back to a cable network.

While we'll start with some examples of ground-level pieces of network infrastructure, this section also covers what might be called networked infrastructure—objects that receive or transmit data across a network but aren't connected to or accessible via the public Internet. These objects are mainly used for city services. Finally, this section includes a very brief list of landmark buildings for New York City's network infrastructure. Because of security concerns, you probably won't be able to access the cool infrastructure parts of the buildings, but in at least a few of them the lobbies alone are worth checking out. These are also good starting points for beginner infrastructure-sightseers to train themselves to search for infrastructure on the street. Since these buildings hold major concentrations of fiber, their perimeters tend to have a lot of orange spray-paint markings and relevant manhole covers.

JUNCTION BOXES

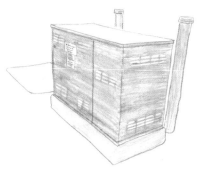

For objects that are so bulky and obtrusive, it's surprising how easy it is to miss these gray and green boxes on the street. More common in outer boroughs than in Manhattan, these boxes are basically the ground-level switching stations for home cable connections. Within these boxes are thousands of wires and cables for telephone, television, and the Internet, all coming from nearby buildings. In the junction box, those cables get connected to terminals that are themselves spliced into the underground cable network.

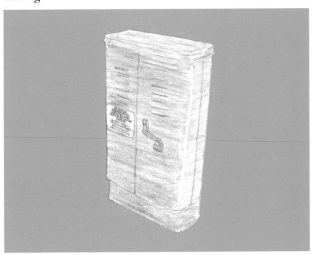

PEOPLE WORKING IN OPEN TELECOMMUNICATIONS MANHOLES

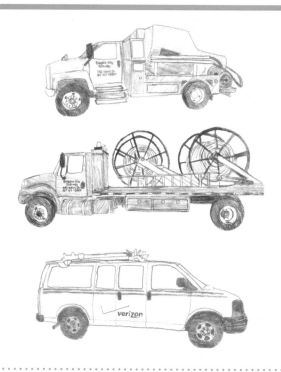

Most of the time, construction or street excavation work is something that people in most cities try to walk around rather than stop to look at. Good indicators of whether the work happening at a particular site is tele-communications-related are the types of vehicles sur-rounding the site and the kinds of equipment and cables visible. It's also helpful to look for certain company

names. (Verizon, Empire City Subway, and Hugh O'Kane Company are among the companies most commonly seen working in Manhattan ducts.)

Sometimes it's possible to take a peek inside an open manhole or handhole to see what's going on under the street. In open manholes, you'll often see large cylinders into which a whole bunch of cables feed in and maybe only one cable feeds out. These cylinders are fiber splice enclosures, where different fiber optic cables get spliced into another cable. In handholes like the one illustrated here, you'll sometimes see devices that connect and convert signal from coaxial cable into optical cable and feed older coaxial from buildings into a fiber network.

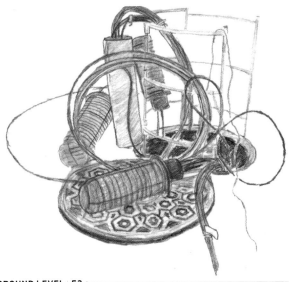

Hugh O'Kane Electric Company

Founded in 1946 as a general electric and maintenance contractor, Hugh O'Kane Electric Company is now one of the top independent fiber installation contractors in New York City. It pulls and splices cable for most of the major networks in the city. In 2002, the O'Kane family created Lexent Inc., which owned and operated dark fiber services company Lexent Metro Connect until 2010, when the company was sold to Lightower. The O'Kane family has apparently continued to work in the fiber leasing world, and many former Lexent employees work for network services startup ZenFi, which shares an office address with Hugh O'Kane Electric Company.

NYCWIN

The New York City Wireless Network (NYCWiN) is a citywide broadband wireless network project initially proposed in 2004 for emergency first responders. While NYCWiN itself isn't necessarily easy to "see" since it's mostly a bunch of cell towers comprising a wireless network, the actual impact of NYCWiN is pretty visible through certain devices connected to its network, including the next two devices described in this guide.

Construction of the network began in 2006 under a $500 million contract with defense contractor Northrop Grumman ($20 million of which came from a Department of Homeland Security grant), and the network became operational in 2009. Some regard the project as a failure given its relatively limited use by city agencies (according to the *New York Daily News* in 2012, less than 15 percent of the network capacity is used on a daily basis) and its exorbitant cost (around $40 million annually just to maintain).

In 2015, the city announced a Request for Expression of Interest and Information (REOI) seeking potential vendors to take over NYCWiN operations. Essentially the vendor would buy the network from the city and then resell municipal services on the network (public Wi-Fi, city agency services, etc.) back to the city. As of this writing, it's unclear what, if any, vendor would take over NYCWiN from the city.

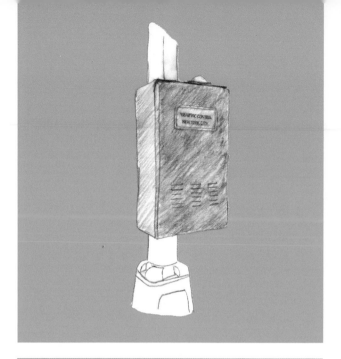

TRAFFIC SIGNAL CONTROLLERS

When first introduced in the 1950s, traffic signals operated electromechanically, using simple timers that changed traffic lights at fixed intervals. Over time, these systems became computerized and networked. Given the size and complexity of New York City's traffic network, it makes sense it would develop an equally massive and complex system of sensors and networked objects to control it.

The dark green signal control boxes attached to traffic signal posts throughout New York are just one piece of a massive system of networked objects. The system, designed by the Nashville–based transit services company TransCore, combines data collected by real-time traffic cameras, RFID (radio frequency identification) scanners, and other field sensors to create traffic signal times that adapt to the immediate conditions of traffic. Each signal control box contains wireless routing equipment and traffic controllers that connect back to a fiber hub. The little green dome on top of the signal control is actually a powerful wireless router used for communicating with the other sensors in the traffic network and the city's Traffic Management Center in Long Island City. Initially piloted in 2011 and slowly rolled out to New York City's over 12,500 traffic signals, this system couldn't have really come to fruition without NYCWiN, which provides the communications backbone that enables all these pieces of the traffic system to talk to one another.

AUTOMATED WATER METERS

Aside from the Department of Transportation, the other major user of NYCWiN is the Department of Environmental Protection (DEP), who implemented a network of automated water meters in 2008. The meters are connected to low-power radio transmitters that send meter readings to NYCWiN antennae on city rooftops, which in turn send those readings to the DEP's servers. The readings are used to verify billing for water use and to detect potential leaks.

MOBILE
LICENSE PLATE READERS

There are more than 34,000 police officers in New York City and more than 8,000 police vehicles. While police cars were networked to each other long before the Internet thanks to radio communications, in the last few years the NYPD has pursued increasingly impressive networked tools to help cops do their jobs.

Starting around 2006, NYPD began equipping some NYPD cars with Automated License Plate Readers (APLRs), devices that photograph and store records of license plates of vehicles on the street. The APLRs on NYPD vehicles are manufactured by ELSAG North America, a subsidiary of Italian company Finmeccanica. Its Mobile Plate Hunter-900 can capture up to 1,800 license plate reads per minute. The camera takes a picture of a passing vehicle's plate and then processes that image into raw letters and numbers that feed into a central database maintained by the NYPD. These plate

records, which include the location, date, and time that the plate was captured, are kept in NYPD databases for five years.

While law enforcement tends to point to the usefulness of ALPRs in tracking down stolen vehicles, the NYPD's first foray into the technology began as part of what was then called the Lower Manhattan Security Initiative, a post-9/11 project that initially focused on security for the Financial District and later expanded to include Midtown and then the rest of the city. In other cities throughout the United States, state and municipal police departments have also acquired ALPRs in the service of counterterrorism or security, as federal agencies like the Department of Homeland Security, the Drug Enforcement Agency, and Customs and Border Protection offer grants to help police departments purchase this type of technology. The majority of vehicles I've seen with ALPRs are marked as CTB—Counterterrorism Bureau.

ALPRs aren't only vehicle-mounted; the devices are also placed at street intersections and bridge and tunnel entrances, sometimes used in conjunction with speed cameras. As of 2013, the NYPD had fewer than 400 ALPR devices, some of which were vehicle-mounted and some of which were not. And that same year, the NYPD collected more than 16 million records of license plate data.

In 2015, the NYPD signed a $442,500 contract with license plate reader company Vigilant Solutions for a sub-scription to their ALPR database, which boasts 2.2 billion records of nationwide license plate data. Since its founding in 2005, Vigilant built up its database by selling ALPRs to vehicle recovery and repossession companies, which would passively collect license plate data as company tow trucks drove throughout a city and then send that data back to Vigilant. The company has a pretty strict terms of service policy that prevents police departments from discussing the program or its services with the media, so there's little public information about what the NYPD is doing with these records aside from the initial announce-ment of the contract.

LINKNYC: FUTURE
NETWORKS OF NEW YORK

In 2012, Mayor Michael Bloomberg announced the Reinvent Payphones Design Challenge, a competition seeking proposals for new initiatives to replace the city's thousands of unused and sometimes unusable public payphones with new technology that could be more accessible and functional for today's city residents. After a lengthy review process, the city announced in 2014 the selection of a proposal called LinkNYC, a network of free wireless hotspots throughout the city. The company behind LinkNYC, CityBridge, was actually a consortium of four companies (Titan, Control Group, Qualcomm, and Comark) that specialized in various facets of the project. The "links" are supposed to offer free Wi-Fi, a touchscreen tablet with maps and other useful local information, domestic phone calls, and charging stations for mobile devices. They also offer advertising space, which is how LinkNYC intends to cover the cost of the service.

While using advertising to support municipal services isn't a radically new idea in New York (just look at the subway system!), LinkNYC's reliance on advertising revenue has raised concerns that the initiative may only further increase the city's existing digital divides rather than decrease them. After the project's initial announcement, the *New York Daily News* reported that the connection speeds offered on LinkNYC kiosks without advertising—primarily kiosks in lower-income neighborhoods—would be much slower than the speeds available on the kiosks with advertising. While the city has argued that this tiered system is temporary and still better than nothing, it is unclear whether advertising revenue will be enough to cover the costs required to bring high-speed fiber cables into neighborhoods that currently don't have them.

Similar to the negotiations that led Empire City Subway to become a subsidiary of Verizon, LinkNYC's consortium has also been reshaped by that familiar alchemy of mergers and acquisitions. In between the initial announcement of LinkNYC and the installation of its first test nodes in the East Village in winter 2015, Google (now Alphabet) subsidiary Sidewalk Labs acquired and merged two of the major companies working on LinkNYC: Titan (the franchise holder for most of the city's existing pay phones) and Control Group (the company largely responsible for the functionality of the kiosks, best known for its work on the MTA subway system's information touch screens). This is one way of saying that the mega-corporation behind Google now has a small but significant share of and role in New York City's pilot program to provide ad-supported public Wi-Fi.

LinkNYC is a promising endeavor from the city to bridge local digital divides, though it's far from the first one. In the Brooklyn neighborhood of Red Hook, local nonprofit Red Hook Initiative has been operating a local wireless mesh network, Red Hook Wi-Fi, since 2011, and in 2013 the Bloomberg administration rolled out a free Wi-Fi network in Harlem designed to cover ninety-five square blocks. LinkNYC's chief distinctions are its installation of a custom hardware unit (the actual "link" kiosk), which appropriates the existing network infrastructure of the telephone grid, and its particular brand of ad-supported public-private partnership.

As of this writing, 134 links have been installed

throughout the city. While the program has mostly been lauded for its ambition and technical achievements, the New York Civil Liberties Union has raised concerns about possible risks faced by LinkNYC users based on the terms of CityBridge's privacy policy. The policy states that CityBridge collects a tremendous amount of user data (including information about a user's device and their on-line activity) that may be used for a variety of purposes, from technical administrative applications to "[providing users] with information about goods or services that may interest [them]"—an indication that the ad-supported service will seek far greater granularity in its targeting than a mere street-level banner advertisement. The NYCLU has also raised concerns about CityBridge sharing user data with law enforcement. While all of these concerns are extremely valid, the fact that LinkNYC is a private company makes it difficult to hold them accountable—they have no public service prerogative to protect or delete user data, and every business incentive to collect and keep it. To paraphrase George Orwell, if you want a vision of the future of public Wi-Fi, imagine a corporation doing exactly the kind of vaguely slimy things corporations do by design—forever.

CARRIER HOTELS AND DATA CENTERS: ARCHITECTURE FOR THE INTERNET

Sometimes people who want to learn about seeing Internet infrastructure ask me to tell them "where the Internet lives." At first glance, this seems like a bit of a misnomer—the Internet isn't a static object, it's defined by the constant movement of information. It doesn't "live" anywhere; it's already everywhere at once—it "lives" in the library down the street, in office buildings, in undersea cables. But there are a few specific types of buildings that hold crucial pieces of Internet infrastructure—less homes for the Internet than waystations that data traffics through. While we'll look a bit at data centers in this section, the buildings we'll primarily focus on are often called "carrier hotels" because it's sort of where different ISPs and network companies "check in" with one another.

Imagine someone sitting at home trying to watch something on Netflix. They click on a movie they want to watch and that click sends a request to Netflix's servers saying, "Hey, bring me the movie *Terminator 2: Judgment Day*!" The person watching Netflix is connected to the Internet via Company A, and Netflix is connected to the Internet via Company B. At some point, the request for *T2* has to move from the Company A network to the Company B network, and carrier hotels are where it happens. Racks and racks of switching equipment and cables run through these buildings, which are also sometimes called "Internet exchanges" or "meet-me rooms" since it's where networks meet one another.

GEORGE WASHINGTON BRIDGE

HUDSON RIVER

CENTRAL PARK

EAST RIVER

KEY

1. 32 Avenue of the Americas

2. 60 Hudson Street

3. 111 Eighth Avenue

4. 375 Pearl Street

5. 75 Broad Street

6. 195 Broadway

7. 33 Thomas Street

8. 140 West Street

WILLIAMSBURG BRIDGE

MANHATTAN BRIDGE

BROOKLYN BRIDGE

In general, these aren't spaces that are open to the public for tours—in one of them, you're not even allowed to take photos of the lobby. This limited access is typical of major network infrastructure nodes. To some extent this has to do with (valid) security concerns, but from my own experiences attempting to get into these spaces, I suspect the managers of these spaces just don't want to deal with infrastructure tourists. (Unfortunately society does not yet view Internet infrastructure with the same reverence or civic zeal as it does other tourist-worthy infrastructure like the Hoover Dam.) But getting inside the server room, while an exciting experience, isn't necessarily required to appreciate these buildings or their role in the network.

In most of the United States, new infrastructure has a tendency to inherit the landscapes of past infrastructure—Internet cables follow telephone lines, which follow telegraph lines, which follow railroads. New York City's Internet infrastructure is no exception. Although there are other data centers and sites of network exchange throughout New York (I particularly regret not having space for the Staten Island Teleport), this section focuses on buildings around Downtown and Lower Manhattan because of the major role these areas have played, and continue to play, in the history of New York's telecommunications infrastructure.

60 HUDSON STREET AND 32 AVENUE OF THE AMERICAS

New York's Internet history is deeply intertwined with the history of the telegraph and the telephone, and the two buildings that best represent that history are 60 Hudson Street and 32 Avenue of the Americas, which are both located just below Canal Street in Downtown Manhattan.

The stories behind both of these buildings in some ways begin over at 195 Broadway, the original New York

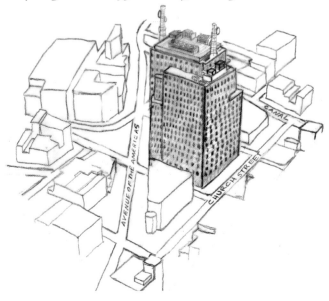

headquarters of the American Telephone and Telegraph Company and Western Union (a space used for offices by AT&T until 1978). In 1914, switching equipment for both companies was moved to 24 Walker Street, but eventually Western Union outgrew this space and in 1928 commissioned the architecture firm Voorhees, Gmelin & Walker to design what would become 60 Hudson Street. In turn, AT&T hired the same firm to create a new building in the same footprint as its Walker Street building and completed 32 Avenue of the Americas in 1932.

Since so much of AT&T's and Western Union's operations overlapped with each other, the creation of the two separate buildings, less than ten blocks apart, also meant the creation of a dense underground cable duct infrastructure. Beneath Church Street, rows and rows of conduit filled with copper wires connected 60 Hudson Street to 32 Avenue of the Americas. As a result, 60 Hudson Street became a major telephone exchange site during the deregulation of the U.S. telephone industry in the 1970s, when nascent competitor telephone companies like MCI and Sprint rapidly moved their equipment into the building in order to take advantage of this duct infrastructure, which made it extremely easy to connect their networks to AT&T's network. 60 Hudson Street's evolution into carrier hotel followed naturally from this period—today, it is home to hundreds of Internet companies' equipment and has the largest concentration of connections to transatlantic cables on the East Coast. 32 Avenue of the Americas's conversion to carrier hotel began with its acquisition by real estate company Rudin Management in 1999.

Not only are both buildings hubs of communication, but they are also magnificent examples of the Deco period in which they were created. Their lobbies harken back to a time when telecommunications had an air of grandeur and idealism (and the wall mosaics at 32 Avenue of the Americas building are an absolute must-see for this).

111 EIGHTH AVENUE

Built in 1932 when Manhattan's ports were far more active in shipping and trade, 111 Eighth Avenue was initially the Port Authority Commerce Building, a warehouse and center for the transport and storage of packaged freight goods, and later became home to some Port Authority offices. In 1998, Taconic Investment Partners turned it into a carrier hotel. In 2010, Google purchased the building for nearly $2 billion. While Google uses a majority of the building for its own office space, the carrier hotel and a number of ISPs, startups, and ground-level retail stores remain. 111 Eighth Avenue is interesting in itself, but it's also a compelling site because of its Chelsea neighbors. Sometimes I think of it as a metaphor for the Internet itself—a weird palimpsest of law enforcement, network infrastructure, spectacle, and commodities.

No building near 111 Eighth Avenue better reflects this idea than Chelsea Market. Formerly part of the National Biscuit Company's factory, the block-long building is now an upscale mall and food court. Restricted-access

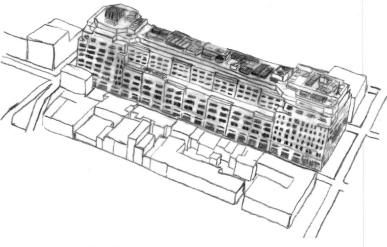

elevators lead to the offices of several cable channels, real estate companies, and other tenants. Google leases three floors of the building too.

Chelsea Market is also home to the NYPD Intelligence Division, which was formed after 9/11 and became notorious for its massively overreaching operations for spying on Muslims. The earliest reference to the existence of the Intelligence Division in Chelsea Market that I found was a redacted NYPD document detailing plans for the 2004 Republican National Convention. The executive summary notes that an "Intelligence Fusion Center" was located in Chelsea Market and served as the "main intelligence gathering and dissemination center" during the Convention. A 2012 document made by Chelsea Market's developer, Jamestown Properties, lists the NYPD as an office tenant occupying 48,000 square feet (for comparison, Google occupies 108,000 square feet in the same building).

Past the Chelsea Market and above the High Line, an enclosed footbridge connects the market to 85 Tenth Avenue, another former National Biscuit Company building turned into a mix of luxury retail, technical infrastructure, and law enforcement. The building is home to a Level 3 colocation center, ground-level expensive restaurants, 360,000 more square feet of Google offices, Möet

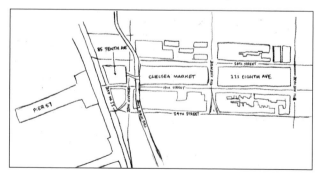

Hennessy's New York offices, and the FBI's Joint Terrorism Task Force.

The JTTF began as a partnership between the NYPD and the FBI in 1980 while investigating the Puerto Rican paramilitary organization Fuerzas Armadas de Liberación Nacional (FALN). Essentially, it's a program designed to make it easier for city police departments and the FBI to work together on investigations rather than having the two agencies work separately on the same case. Today, it has offices in 103 cities; 71 of those offices were created after 9/11.

According to a General Services Administration document from 2014, the JTTF has been at 85 Tenth Avenue since 2005 and intended at that time to renew its lease through 2020 or until it could move to another "government-owned location." It's unclear whether they chose the location for its proximity to Internet cables (Level 3 acquired its colocation space in 1999) or for its raw post-industrial interior, which can accommodate its unusual architectural needs. In *Enemies Within*, a comprehensive volume on the NYPD Intelligence Division, the authors Matt Apuzzo and Adam Goldman describe a "cavernous" secure compartmentalized information facility (SCIF, a fancy acronym for "surveillance-proof government building") on the tenth floor, and a 2015 House of Representatives document approving the lease renewal noted that the task force rented 168,000 square feet at an annual cost of around $13 million. The footbridge above the High Line that connects the building to Chelsea Market is supposedly a direct link between the Intelligence Division and the JTTF, although it remains locked—communication across agencies is apparently Not Their Thing.

One less obviously relevant but still interesting landmark in this accidental luxury-as-cloaking-device tour is located across the West Side Highway from 85 Tenth Avenue: Pier 57, which is currently under development by Youngwoo & Associates to be a luxury retail site re-branded as the SuperPier. The building, a former MTA bus repair center, is more familiar to some as "Guantan-amo on the Hudson" due to its use as an arrest holding site for an estimated 1,200 protesters during the 2004 Republican National Convention.

In a press statement following several settlements to RNC cases in 2014, the National Lawyers Guild described conditions in Pier 57 circa 2004:

> . . . cyclone fencing was used to create cages in a warehouse-like area still covered with grease and brake fluid. Signs still hung from the walls warn-ing workers to wear hazmat suits. There was no heat, no place to lie down, and a handful of port-a-potties. Protesters were held in these dis-graceful conditions for up to 48 hours before be-ing transported to court facilities—long enough to exhaust them and keep them off the streets until after George Bush was re-nominated. Many left with skin rashes and respiratory prob-lems, and some developed more serious medical conditions.

No word as of yet on whether SuperPier tenants Opening Ceremony or Google will incorporate the Guantanamo-on-the-Hudson aesthetic into their interior design.

That an industrial bakery like National Biscuit Company and a Port Authority warehouse would be transformed into a data center and an Internet ex-change is perhaps not a surprise, given infrastructure's tendency to inherit the spaces of preceding technologies (the same could be said for Google's steadily increasing footprint in the area). The presence of high-end retail in former industrial spaces is also a familiar narrative; spaces for other people's leisure love to evoke nostalgia for other people's labor. Law enforcement's placement within this landscape is probably more pragmatic than poetic (raw industrial spaces reworked for retail can be as easily reworked for government-specified security standards), but there is something weirdly dislocating about walking through the retail corridors along Fifteenth Street, aware of the layers of state and infrastructural control a few floors aboveground and layers of fiber-optic networks several meters underground—systems and histories mostly glimpsed by following orange spray-paint markings from Eighth Avenue to the end of Fifteenth Street.

375 PEARL STREET

Relatively younger and decidedly less Deco than other major connection points in Manhattan, 375 Pearl Street was built in 1975 as a switching station for the New York Telephone Company. Taconic Investment Partners (the same company that turned 111 Eighth Avenue into a carrier hotel) purchased the building in 2007 with grandiose plans to transform its much-derided windowless exterior and add new office space and condominiums. The economic collapse of 2008 pretty much killed that plan, and 375 Pearl Street ended up being sold at a massive loss to Sabey Data Center Properties in 2011. Sabey rechristened the building Intergate Manhattan, describing it in publicity materials as "the world's tallest data center."

As with 111 Eighth Avenue, 375 Pearl Street is consid-

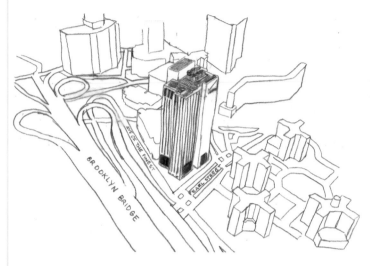

ered remarkable in part due to what's around it, or more specifically what surrounds it. Its next-door neighbor is 1 Police Plaza, the NYPD's headquarters, so the building has police checkpoints on almost every side.

75 BROAD STREET

There are a few small indicators that this building in the heart of Manhattan's Financial District was once the heart of a major telecommunications company. Built in

1929, 75 Broad Street was originally the headquarters of the International Telephone and Telegraph Company (ITT), a telecommunications conglomerate founded in 1920 through the acquisition of various Caribbean and European communications companies. ITT established its U.S. presence by acquiring industrialist John William Mackay's various telecommunications ventures: the Commercial Cable Company, the Commercial Pacific Cable Company, Postal Telegraph, and the Federal Telegraph Company. The building entrance at the corner of Broad and William Streets features a mosaic depicting an angel foregrounded by maps of the Eastern and Western Hemispheres, apparently connecting the two sides of the world with the wonder of communications technology (depicted here as lightning bolts). ITT's colorful history may be too voluminous of a detour within this guide (highlights include: collaboration with the Nazi party and

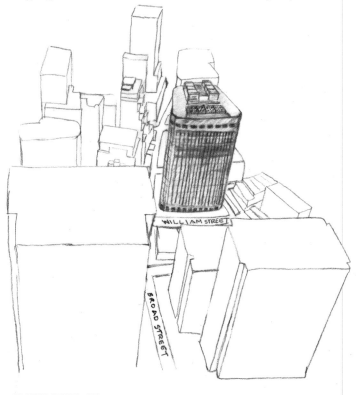

Nazi-sympathetic governments, working with the CIA to covertly finance the 1973 coup of Chilean president Salvador Allende, being bombed by the leftist radical organization the Weather Underground for involvement in the Chilean coup, and being the subject of the Fela Kuti song "International Thief Thief"—seriously, this company was really evil). In any case, it had left 75 Broad Street by 1961 and sold off its telecommunications assets to what would become Alcatel-Lucent in 1986. It wasn't until 1999 that Newmark and Company repurposed several floors of the building into a data center.

75 Broad Street's central Lower Manhattan location, while great for its proximity to Manhattan carrier hotels, proved terrible in 2012 when Hurricane Sandy hit the New York City area. Although the data centers had backup generators, the building's data center operations were on the eighteenth floor, which meant that operations managers had to carry fuel up eighteen flights of stairs while the power was out throughout the Financial District.

While some of these buildings have little or no remaining telecommunications equipment, they are notable landmarks of New York's network history, worth checking out if you have the chance to do so.

195 Broadway

The original headquarters of AT&T and Western Union prior to the creation of 60 Hudson Street and 32 Avenue of the Americas. It was used as AT&T's offices between 1916 and 1978, and in 1927, the first transatlantic telephone call—between London and New York—took place here.

33 Thomas Street

Another AT&T Long Lines building, 33 Thomas Street, was completed in 1974. It was and still is mostly used for telephony services, but it's both a Brutalist monstrosity of a building and highly visible on the horizon from both 60 Hudson Street and 32 Avenue of the Americas, so it bears a quick mention.

140 West Street

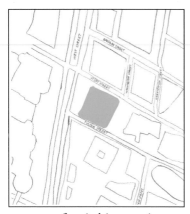

Constructed in 1926 by the same architects as 60 Hudson Street and 32 Avenue of the Americas, 140 West Street was the headquarters of the New York Telephone Company, the predecessor company of Bell Atlantic and, later, Verizon. While today it only has a limited amount of switching equipment, it's notable in part because of the damage it incurred during, and repairs made after, the 9/11 attacks.

ABOVE GROUND

In this section, we'll look at objects that tend to be above eye level—mostly wireless devices transmitting signals across networks. Some of these networks are the hardest to see, since they either reside on rooftops or above traffic intersections (which, in New York, aren't exactly a good place to linger). But as more and more Internet usage moves to wireless devices like smartphones and tablets and as more and more wireless devices become part of the regulation and management of citywide logistics and public safety, wireless infrastructure becomes increasingly crucial to understanding how people live with the Internet—with networked objects in general—in cities.

Some of the networks in this section are not the obvious ones typical users "connect to" regularly, like the public Internet. Surveillance cameras are perhaps one of the most noticeable—and contested—examples of non-Internet networked objects in public space. Traffic sensors are another. The public's "connection" to these networks is admittedly more oblique than connecting to a cell tower, but we do connect with them frequently simply by engaging with and in public space.

CELL TOWERS

When out and about on the street, more and more people connect to the Internet through wireless networks. For many, seeing the Internet on the street just means using a smartphone. Since those wireless operations happen at a level invisible to the human eye, it might be helpful to explain what exactly is going on when phones connect to cell networks.

To start with: cell phones are basically screaming all the time. We can't hear them screaming because they scream in radio waves, but they're constantly announcing their existence to other antennae via these radio signals. They're not saying all that much most of the time—more or less just "Hi! I'm here! I'm looking for a network to connect to!"

If the phone is close to a cellular tower with an antenna that connects to that phone's particular carrier, it connects to the network via that antenna. If the phone moves away from that antenna or gets out of range of it,

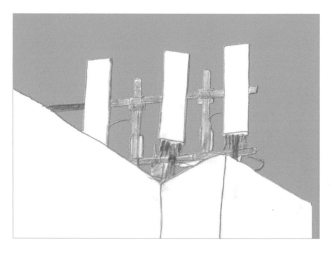

that's okay—the phone is still screaming. The next nearest antenna will pick up its signal.

After someone dials a number or opens an app on their phone, the cell phone sends a signal out to the nearest tower with the request for that call or that app (it's still screaming "Hi! I'm here!" but now also screaming "Bring me this app!" or "Call this person!"). The cellular antenna receives the request and sends it back into a fiber optic cable network, routing it through a much larger network (either of more cables or microwave antennae) and to the right server that can process the request (e.g., a call switching station or a Facebook data center). That server sends the requested data back through the network to the antenna *nearest* to the phone and that antenna sends the data back to the phone.

In New York, cell towers are generally pretty hard to see from the street—they're mostly on the tops of buildings. They're also often disguised, although their New York disguises (bricks on buildings) are pretty simple compared to cell tower disguises in other places (trees, church crosses, cacti).

MICROWAVE ANTENNAE

While these antennae are damn near impossible to see most of the time from the street, it's useful to know about their existence. There are a small handful of wireless Internet service providers (WISPs) in New York City, who provide broadband services, mainly to businesses, through a network of antennae. At least one of these antennae is usually on the roof of a major carrier hotel, which is where data gets transmitted back into the global Internet. On other rooftops, they'll sometimes be alongside or attached to other antennae, as in the illustration above featuring a small microwave antenna affixed to a cell tower.

DISTRIBUTED ANTENNA SYSTEMS

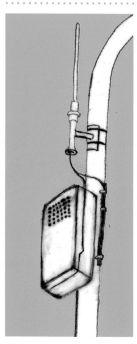

A Distributed Antenna System (DAS) is basically a way to expand a cell network's reach, adding capacity in under-covered areas. They're a little easier to find on the street because they're not on top of buildings—they're attached to street poles and linked to underground fiber optic networks. If you ever see an orange cable marking going into a street pole, look up. You'll probably see a DAS. There are seven companies with franchise agreements to maintain Distributed Antenna Systems in New York; however, three of those companies belong to one company as of 2015 (Crown Castle) and two appear to be subsidiaries of the same company (ExteNet Systems).

PUBLIC WI-FI ROUTERS

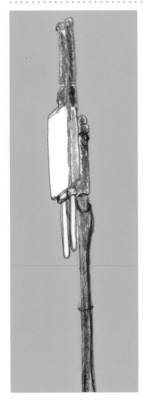

Throughout the city, there are a handful of parks that provide free, public Wi-Fi. This particular illustration is from Madison Square Park. Access to free Wi-Fi in New York City parks is partly brokered by the franchise agreement process. When a company receives a franchise to run cable throughout the city, the agreement usually comes with a caveat that the company has to provide the city with some municipal services and support. Starting in 2011, the New York City Department of Parks began making agreements with local franchisees (first AT&T, then Cablevision and Verizon) to bring Wi-Fi to public parks. The resulting rollout and reliability of that Wi-Fi has been pretty spotty to date, but even in parks where the Wi-Fi isn't readily available (such as the park where I spotted this router), artifacts of those networks remain.

RFID E-ZPASS READERS

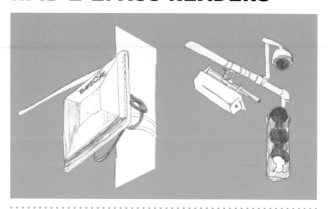

Most intersections in New York City have either one of these two antennae devices. Both devices broadcast and receive radio signals. In this context, they're used to read radio frequency identification (RFID) devices embedded in E-ZPass devices. Technically E-ZPass is used for toll collection, but E-ZPass Readers and other sensor devices also collect data from RFIDs for traffic monitoring purposes.

E-ZPasses work by registering drivers' travel through toll booths via the transponder, another word for the RFID that drivers keep in their car. When the E-ZPass is in proximity of an antenna that can pick up the transponder's frequency, the transponder transmits uniquely identifiable information to the antenna (like an E-ZPass account number). Once the antenna at a toll booth receives this information from the transponder, it relays that information back into a larger network, which is where the actual E-ZPass payment processing happens.

The E-ZPass readers above intersections in New York City aren't for toll collection at all. They're exclusively intended to monitor traffic patterns, measuring the number of cars passing a given intersection or road and adjusting traffic light patterns accordingly. Originally part of the "Midtown in Motion" smart traffic pilot project, these sensors are now present in many other parts of the city. A Freedom of Information request by the New York Civil Liberties Union didn't uncover how long data from the E-ZPass readers is stored or whether it remained only in the city's possession.

MICROWAVE RADARS

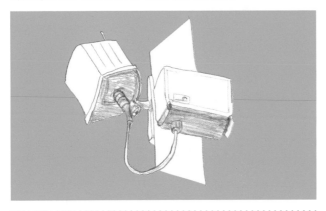

Another original component of the "Midtown in Motion" project, Remote Traffic Microwave Sensors (RTMS) are now used in many other parts of the city and are popular with transit agencies throughout the country as a low-cost, low-maintenance method of counting and tracking traffic in intersections. The RTMS detects motion and speed by measuring the distance of objects in its microwave beam's line of sight. Presumably these traffic sensors compensate for the number of non-E-ZPass-equipped vehicles that aren't picked up by the RFID readers.

SHOTSPOTTER

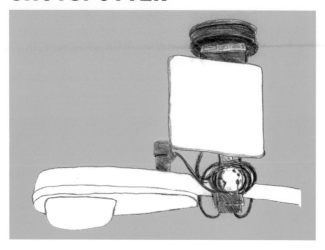

ShotSpotter is a company based in Newark, CA, that produces acoustic sensor technologies used to detect gunshots in city streets. The sensors are equipped with a microphone, GPS, and some processing and wireless transmission capabilities. When three or more sensors detect a noise that might be a gunshot, data from the sensors is transmitted to ShotSpotter's Incident Review Center in California, where analysts review the waveform pattern of the audio collected and listen to verify if the sound is, in fact, a gunshot and not something like fireworks or a car backfiring. The rationale for a sensor network to detect gunshots is essentially that many people don't report gunfire to 911.

In July 2014, the New York Police Department entered into a $1.5 million contract with ShotSpotter to operate a pilot program using the technology in the Bronx;

in March 2015, the program expanded to Brooklyn. The 2016 city budget allocated $1.8 million for the fiscal year and $2.5 million in 2017 to expand the sensor network from seventeen precincts to forty-five over two years.

Although the technology is used in more and more cities throughout the United States, ShotSpotter is not without controversy. In a 2012 case in New Bedford, MA, audio recorded following gunshots was used to identify a suspect in a murder case—which was the first indication that ShotSpotter's sensors could, in fact, record and store audio that wasn't necessarily from a gunshot. Some cities that have tried ShotSpotter, such as Trenton, NJ, and New Haven, CT, have questioned the effectiveness of the technology given the high rate of "false positives"—i.e., loud noises identified as gunshots and no indication of gunfire when police arrive on the scene. As of this writing, the NYPD hasn't released any updates on the success of the program and a bill introduced by Public Advocate Letitia James requiring the NYPD to publicly release ShotSpotter data has lingered in committee for over a year.

As far as networked devices go, ShotSpotter's sensors are tricky because they don't call that much attention to themselves. It would be easy to mistake a cluster of antennae and cables for some other traffic sensor or perhaps a part of the NYPD's camera network. And it's possible that they will become even more difficult to identify: the same year that the NYPD expanded its use of ShotSpotter to Brooklyn, the company entered into a partnership with General Electric Lighting to develop embedded gunshot detection sensors for GE's intelligent LED street light fixtures, which also monitor weather and traffic conditions.

SURVEILLANCE CAMERAS

TRAFFIC CAMERAS

The New York City Department of Transportation (DOT) and the Metropolitan Transit Authority (MTA) operate traffic cameras at, respectively, 723 intersections and 20 bridge and tunnel entrances. These cameras are used for traffic monitoring purposes. The DOT cameras and MTA cameras both have live streams of their footage available online.

MTA SUBWAY CAMERAS

It's unclear, from what I've been able to find, exactly when the MTA began installing closed-circuit television cameras on some subway platforms, but efforts to expand that camera network ramped up dramatically after September 11, 2001. The MTA currently has more than 4,500 cameras operating throughout the transit system, with 1,500 of those cameras on city buses. Data collected by cameras feeds back to MTA Rail Control Center, located on 54th Street between Eighth and Ninth Avenues in Manhattan.

Following 9/11, security became a major priority for city agencies, and the MTA was no exception. Its 2000–2004 budget allocated $591 million for security projects, and in 2005 the agency issued a $212 million contract to defense contractor Lockheed Martin to provide a state-of-the-art security system for the agency. The system was to

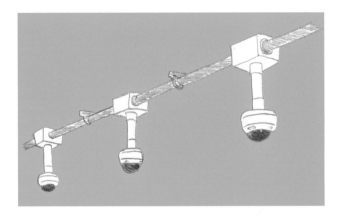

include 3,000 networked cameras and a network of sensors to identify suspicious packages or objects.

However, some of Lockheed's high-tech promises never really came to fruition, and in 2009 the contractor found working within the bureaucracies of the MTA so onerous that it sued to get out of its contract with the city. The MTA filed a countersuit shortly thereafter.

The MTA-Lockheed lawsuit couldn't have come at a worse time in the agency's history—by 2010, the MTA's finances were in such disarray that the agency ultimately had to cut services and institute its now-biennial fare increases. A 2010 article about the lawsuit noted that the $3.6 million the MTA had already spent in litigation was equivalent to the cost of the ten bus lines the agency planned to cut. The lawsuit remains in litigation as of this writing. New contractors continue to work on the electronic surveillance network.

CRIMEEYE CAMERAS

These cameras are sort of rare finds—they're visible mainly in Lower Manhattan and pretty much only around federal buildings. They appear to belong to the Department of Homeland Security and are manufactured by a company based in Suffern, NY, called Total Recall Corporation. (Not even joking, that's their name.)

NYPD CAMERAS

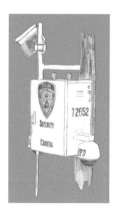

The New York City Police Department has a few thousand white, labeled surveillance cameras that, according to press reports, are part of a program called Argus. In Greek mythology, Argus was the name of a giant with one hundred eyes. Apparently, coming up with a clever name for a surveillance tool is really hard, so when searching for information about the NYPD's Argus, one pretty quickly finds other surveillance camera products with the same name and police departments calling *their* new exciting initiative Argus.

Earliest reference to the program dates back to 2006, when there was an initial install of five hundred of the cameras in the city. At the time, press reports noted one

interesting distinction about the new cameras: they communicate wirelessly. This communication happens via the white rectangular patch antenna attached on top of the white box. If you see one of these white patch antennae on top of an NYPD camera, look around the roofs of nearby buildings and other lamp posts in the area—chances are, you'll find another antenna. These antennae form a point-to-point wireless system, in which information from one device (in this case, a surveillance camera) travels wirelessly from its antenna to another node within its line of sight, at which point it's transmitted back to a wired network and some central location. While there are some wired cameras in the NYPD's network that were installed prior to 2006, wireless surveillance cameras offer the advantage of not requiring the installation or splicing of new cables for every new camera installed.

The antennae used on the NYPD's cameras are a product of Proxim Wireless, a company that makes wireless broadband networking systems primarily for large-scale, outdoor applications in business or municipal government contexts. The Argus camera system as a whole, however, is not built by Proxim.

These are another make and model of NYPD camera sometimes seen around the city. It's unclear what, if anything, distinguishes them from the white-boxed cameras.

The Domain Awareness System

Through tracking the installation of NYPD cameras through press reports and city council records approving new cameras, a piecemeal portrait of the city's camera network emerges, but getting a big-picture overview of the entire network is pretty difficult. The NYPD would probably prefer to keep it that way. When I filed a Freedom of Information Act request for the exact number and locations of these cameras, I was denied on the grounds that it would reveal "non-routine techniques and procedures"; furthermore, disclosure "would enable the planning of criminal activity so as to reduce the possibility of being caught on video."

Attempts by the public to track, count, or map surveillance cameras (police-owned or otherwise) have in general been pretty unsuccessful. Part of the problem of mapping out police surveillance cameras is the sheer scale of the network and the difficulty of organizing enough people to do the counting. But the other problem is that such an undertaking will always be incomplete, as it only reflects cameras that have been *labeled* by the NYPD. It doesn't begin to factor in the secret, unmarked cameras, the privately owned cameras that individuals readily turn over to police, or the privately owned cameras that feed directly into the NYPD's existing citywide surveillance network, the Domain Awareness System.

Not to be confused with Distributed Antenna Systems, the Domain Awareness System (also DAS) is the city's massive counterterrorism apparatus that collects and analyzes all of the information from police-operated networked devices previously mentioned in this field guide. Built in collaboration with Microsoft in 2012, the DAS allows police to connect content from camera feeds with arrest records,

911 calls, and license plate recognition technology. Under the terms arranged with Microsoft, New York receives a 30 percent cut of any sales Microsoft makes of DAS software to other cities. As of this writing, the most recent documentation I could find about how much money the department has made from this arrangement was a 2015 *Wall Street Journal* story, which noted that NYPD deputy commissioner of information and technology Jessica Tisch had framed and hung in her office the first of the checks from this profit-sharing arrangement, which amounted to $375,355.20 (although, for context, the NYPD's annual budget as of fiscal year 2016 is $4.8 billion).

The DAS isn't exactly a brand-new endeavor; it's more the current incarnation of years of post-9/11 initiatives to increase security in New York City. Some of the initial groundwork for this system dates back to the 2004 Republican National Convention in New York City, during which new cameras and the NYPD's existing Emergency Operations Center received major technical improvements and upgrades. In 2005, the NYPD launched the Lower Manhattan Security Initiative (LMSI), a project to tighten security specifically around Lower Manhattan similar to London's "Ring of Steel." The program was initially funded with $10 million from the Department of Homeland Security (DHS) and $15 million from the city and involved some contracting with IBM. The LMSI expanded its surveillance coverage and became the Midtown Manhattan Security Initiative in 2009 (with costs cited somewhere around $201 million, of which approximately 90 percent came from DHS). Estimates for the costs of the Microsoft partnership, created in 2012, range between $30 and $40 million. A number of the system's cameras belong to private "stakeholders" including the Federal Reserve, Goldman Sachs, and Pfizer, who have access to the DAS headquarters at 55 Broadway, an office building at the corner of Exchange Place.

APPENDIX: TYPES OF CABLE

COPPER: Originally used for telegraph and telephone communications starting in the 1880s, a surprisingly large amount of copper remains in New York City's ducts despite severe damage incurred during Hurricane Sandy. This is in part because New York is an old city and no one wants to inadvertently cut a phone line presumed dead. It's also a shrewd real estate trick by incumbent telecoms—opening up space in the ducts by removing dead cable means a competitor can put new cable in.

COAXIAL: A coax cable still has a copper conductor, but is more insulated and more efficient than older copper wires. It's commonly used for cable television connections. A lot of residential buildings in New York still rely on coaxial connections, which connect to fiber networks in utility cabinets on the street or underground.

FIBER: Optical fibers are transparent strands of glass that transmit data as pulses of light. It offers far faster speeds and greater bandwidth for transmitting information than copper and coaxial. While a part of large-scale telecommunications infrastructure since the 1970s, fiber connections to residential buildings (the so-called "last mile" connection) have only really emerged in the last decade or so, partly due to consumer demand for higher speeds and partly due to the dropping cost of fiber itself.

DARK FIBER: Companies that receive a DoITT franchise can sell or lease unused fiber (known as dark fiber, as in not "lit" with data transmissions) to other businesses. This glut of unused fiber is a consequence of the first dot-com bubble, when lots and lots of ISPs that were building out networks collapsed as the bubble burst. These collapsed ISPs and their networks were eventually absorbed by larger companies. Dark fiber is typically leased for private use by organizations that want exclusive network connections for either speed or security reasons, such as academic institutions, banks, and government agencies.

GLOSSARY OF TERMS

NOTE: This section is and will always inevitably be incomplete; an exhaustive list of relevant terms and concepts is itself a separate book entirely. If only there were some technology or service that organized and rendered information accessible that readers might be able to turn to should they not recognize a term used in this field guide!

ANTENNA: a device used to transmit or receive radio waves.

AUTOMATED LICENSE PLATE READER (ALPR): a device that photographs license plates and translates the image of the plate into raw letters and numbers to be stored as machine-readable data.

CARRIER HOTEL OR INTERNET EXCHANGE: a location where Internet service companies place equipment that allows them to transfer Internet traffic from their network to other companies' networks.

DISTRIBUTED ANTENNA SYSTEM (DAS): a network of objects that can transmit wireless signals (typically low-power antennae), most commonly used to create a more reliable signal across either an indoor or outdoor space.

DoITT: New York City's Department of Information Technology and Telecommunications, which is responsible for the maintenance of the city's information technologies and telecommunications infrastructure.

FIBER SPLICE ENCLOSURE: a container that holds different fiber optic cables that need to be fused together.

FRANCHISE AGREEMENT: in New York telecommunications, this is an agreement between the city and a telecommunications company or service provider that allows the provider to install and maintain cables or wireless devices in the city and to provide services to New York residents.

INTERNET: a network composed of different networks communicating with each other using Transmission Control Protocol and Internet Protocol (TCP/IP).

INTERNET SERVICE PROVIDER (ISP): a company that provides Internet access.

JUNCTION BOX: typically, a small cabinet on the street that houses cabling and equipment for transferring communication signals from nearby buildings into an underground cable network.

LINE OF SIGHT: in wireless communications, the distance between two antennae communicating with each other without obstruction. Things that can obstruct line of sight in wireless communications include buildings, trees, and (on occasion) dense cloud cover.

MICROWAVE: a form of electromagnetic radiation that emits waves that can be used for communications purposes (among other uses).

NETWORK: in this context, a system that permits computers to exchange data with each other.

PEERING: the creation of an interconnection between separate networks in order to exchange information (in the case of Internet networks: traffic) between them.

POINT-TO-POINT WIRELESS: a type of communications network in which information travels wirelessly from one device to another.

RADAR: a system that uses radio waves to identify the position and speed of objects moving through space. Originally an acronym created by the United States Navy (Radio Detection and Ranging).

RADIO FREQUENCY IDENTIFICATION (RFID): a means of using electromagnetic fields to read and process electronically stored information attached to objects.

ROUTER: a device that transfers information across networks.

SENSOR: objects that typically use electrical or optical means to detect changes in an environment.

UTILITY DUCT: an underground (or, sometimes, aboveground) conduit for housing cables or pipes for utilities such as power, water, gas, or telecommunications.

WI-FI: a wireless technology that allows computers to communicate via either 2.4 GHz or 5 GHz radio. Technically the actual technology involved creates a wireless local area network (WLAN); the term Wi-Fi has become commonplace through the weird fluke of marketing and Internet history. The word Wi-Fi actually means nothing.

WIRELESS INTERNET SERVICE PROVIDER (WISP): a company that provides Internet access and services specifically using wireless networking technology.

ACKNOWLEDGMENTS

Thanks to the following people for their assistance in the creation of this book. Some of them did a lot of useful things; some of them did one tiny but crucial thing. They're all great and I appreciate them very much.

- Ross Andersen
- Andrew Blum
- James Bridle
- Aaron Straup Cope
- Kate Crawford
- Marina Drukman
- Neil Freeman
- Sha Hwang
- Erin Kissane
- Sam Kronick
- Charlie Loyd
- Tim Maly
- Surya Mattu
- Jonathan Minard
- Juna Mercury
- Robinson Meyer
- Mayo Nissen
- Peter Richardson
- Eleanor Saitta
- Taylor Sperry
- Allen Tan
- Dan Williams

Thanks also to Data and Society Research Institute, Eyebeam Art and Technology Center, the School for Poetic Computation, and, of course, Melville House.

ABOUT THE AUTHOR

INGRID BURRINGTON writes about the Internet, politics, and art, and has been published in *The Atlantic*, *The Nation*, *ProPublica*, *San Francisco Art Quarterly*, *Dissent*, and elsewhere. She's given talks at conferences both in the United States and abroad, and her art has been exhibited in galleries in New York, Tokyo, Leipzig, Baltimore, Philadelphia, and many other cities. She lives in Brooklyn and @lifewinning.